Tsunamis
Killer Waves

Michele Ingber Drohan

The Rosen Publishing Group's

PowerKids Press™
New York

For Rey—the unique phenomenon in my life.

Published in 1999 by The Rosen Publishing Group, Inc.
29 East 21st Street, New York, NY 10010

First Edition

Book Design: Danielle Primiceri

Photo Credits: pp. 5, 10, 14, 18 © Corbis-Bettmann; p. 6 © Archive Photos; pp. 9, 21 © 1996 PhotoDisc, Inc.

Drohan, Michele Ingber.
    Tsunamis / by Michele Ingber Drohan.
       p.   cm.— (Natural disasters)
    Includes index.
    Summary: Describes tsunamis, where they occur, what causes them, and what can be done to protect people from them.
    ISBN 0-8239-5286-X
    1. Tsunamis—Juvenile literature. [1. Tsunamis.] I. Title. II. Series: Drohan, Michele Ingber. Natural disasters.
  GC221.2.D76 1998
  551.47'024—dc21
                                   98-4579
                                      CIP
                                      AC

Manufactured in the United States of America

Contents

What Is a Tsunami?

A **tsunami** (soo-NAH-mee) is a group of waves that move across the ocean, one after another. These waves are not like normal ocean waves made by the wind. Instead, tsunami waves form because of a **disturbance** (dih-STER-bints) in or near the ocean.

The waves of a tsunami can move as fast as 600 miles per hour. When a tsunami reaches the coast the waves can be as high as 100 feet. Most often a tsunami does not look like a big wave. It looks like a huge wall of water moving forward.

Scientists came up with the term tsunami in 1963. It is a Japanese word; tsu means "harbor" and nami means "wave." ▶

What Causes a Tsunami?

Have you ever heard the term tidal wave? Many people used to call tsunamis tidal waves. But tidal wave is the wrong term. Tsunamis are not affected by the **tides** (TYDZ). Instead, tsunamis are caused by **earthquakes** (ERTH-kwayks) that occur near or on the ocean floor. Most earthquakes happen because the **plates** (PLAYTS) in Earth's **crust** (KRUST) bump into each other. Because the **force** (FORS) of an earthquake is very strong, it pushes large amounts of water to the surface. This is what creates a tsunami. Tsunamis can also form after the **eruption** (ee-RUP-shun) of a volcano, a landslide, or the crash of a **meteorite** (MEE-tee-or-yt) from space.

◄ *Earthquakes don't just cause harm on land. Under the ocean, earthquakes are the major cause of tsunamis.*

What Makes Tsunami Waves Different?

The distance between two waves is called a **wavelength** (WAYV-lenkth). Ocean waves caused by the wind usually have wavelengths of only 30 to 60 feet. But the waves of a tsunami can have wavelengths over 300 miles long. When the tsunami is in deep water the waves are only a few feet high. A tsunami could pass under a ship in the middle of the ocean and no one on the ship would notice it. When the wavelength is very long, a tsunami can travel fast. It can move across a whole ocean in less than a day. But when the tsunami reaches **shallow** (SHA-loh) water, it slows down. The wavelength of the tsunami gets shorter and shorter until one giant wave is formed.

This diagram shows how a tsunami travels across the ocean. ▶

SPEED OF WAVE AT SEA
500 MPH

SPEED NEAR SHORE
80-100 MPH

HEIGHT AT SEA UP TO 3FT

130MI.

HEIGHT NEAR SHORE
CAN REACH 100FT

WAVES SLOW DOWN AS THEY SKID ACROSS THE BOTTOM

A Tsunami's Power

When a tsunami hits land, it can cause a lot of **damage** (DAM-ij) to the land and people. Even though it slows down when it reaches the coast, a tsunami still has a lot of power. In fact, a tsunami is so powerful that it can move large rocks, boats, and cars. A tsunami floods an area and **destroys** (dih-STROYZ) the land. It pulls sand and plant life from the beach. Tsunamis destroy homes and other nearby buildings. Many people don't understand how strong and fast a tsunami can be. Some people have been killed while trying to see a tsunami as it hit shore.

◀ *This house in Seward, Alaska, was crushed by a tsunami following an earthquake in 1964.*

Where Do Tsunamis Happen?

Tsunamis can happen in every ocean in the world. But they form most often in places around and under the Pacific Ocean because many volcanoes and earthquakes occur there. There are so many volcanoes in this area that it has been named the Ring of Fire.

Scientists from different countries around the Ring of Fire have come together to warn people of tsunamis. Japan and Hawaii are part of this system because over the years they have been hit by many tsunamis.

The islands and continents in the Ring of Fire have the most active volcanoes in the world. ▶

RUSSIA

CANADA

CHINA

JAPAN

UNITED STATES

HAWAIIAN ISLANDS

SOUTH AMERICA

PACIFIC OCEAN

AUSTRALIA

RING of FIRE

The Pacific Tsunami Warning Center

The Pacific Tsunami Warning Center (PTWC) was formed in 1965. It watches for earthquakes large enough to cause tsunamis. Earthquakes are measured with **seismographs** (SYZ-muh-grafs). If an earthquake happens, the PTWC sends out a watch. During a tsunami watch, information is given to people by radio and television.

The PTWC looks at the waves and water level near the **epicenter** (EH-pih-sen-ter) of the earthquake to see if a tsunami has formed. If a tsunami has formed, a warning is given. A tsunami warning tells people to **evacuate** (ee-VA-kyoo-ayt) right away!

◀ *This railroad bridge, crossing the Wailuku River in Hawaii, was partially destroyed by a tsunami in 1946.*

Disaster Strikes!

Some tsunamis are caused by earthquakes that occur close by. These are called local tsunamis. Other times, an earthquake happens far away and a tsunami travels hundreds of miles across the ocean.

In 1960 there was an earthquake near Chile. Fifteen minutes after the earthquake happened, a tsunami destroyed the area. Thousands of people died. The tsunami traveled to the other side of the ocean and hit Hawaii fifteen hours later. Sixty-one more people died. Since then, warning systems have helped save people. But even with a warning, it is hard to save people from local tsunamis because they happen so fast.

Tsunamis are unlike other natural disasters because one tsunami can destroy separate ▶ places that are thousands of miles apart.

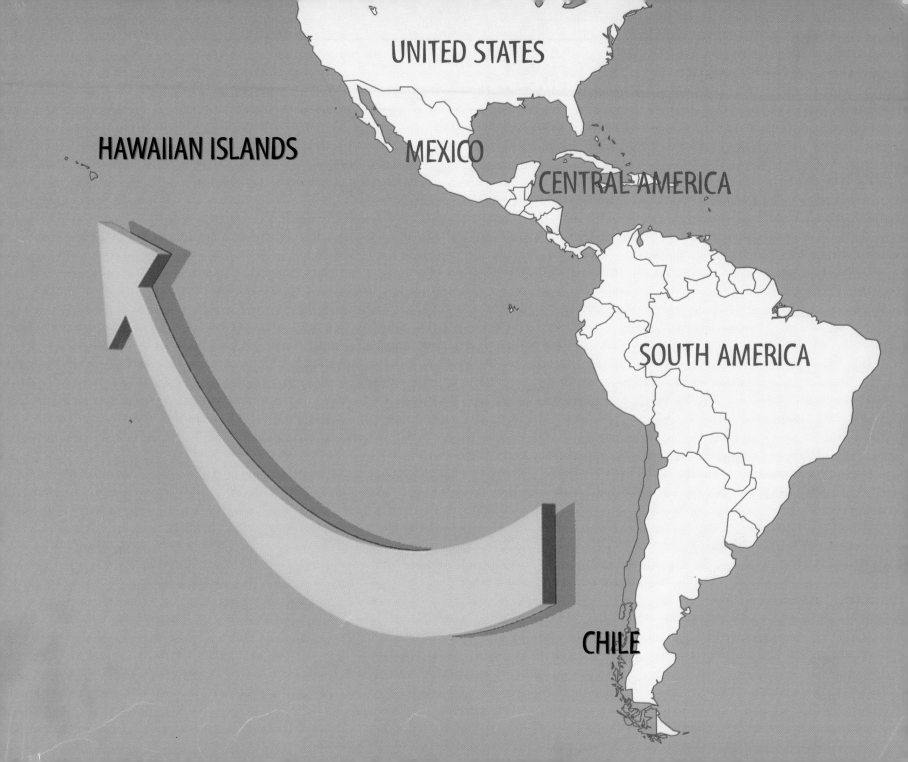

HAWAIIAN ISLANDS

UNITED STATES

MEXICO

CENTRAL AMERICA

SOUTH AMERICA

CHILE

Can You Measure a Tsunami?

No one can measure the height of a tsunami as it is happening. But **experts** (EK-sperts) have found other ways to measure how high the waves are. The **maximum** (MAK-sih-mum) height of a tsunami is called a run-up. Experts find the run-up by looking at the **debris** (duh-BREE) left over from the tsunami. Depending on where the debris is laying, it's possible to figure out how high a wave must have been. Scientists also look at plants and trees that were killed by the salt from the ocean water. Experts do this so they can make evacuation maps. These maps tell people if they are in danger when a tsunami warning is made.

◄ *People should never try to see a tsunami as it's happening. A tsunami moves much faster than a person can run.*

Tsunami Safety

If you hear a tsunami warning, it is very important for you to quickly evacuate your area. Listen to the radio or watch the television to find out what to do and where to go. If the earthquake that caused the tsunami is far away, you will have time to move to a safe place. But if the tsunami is local, you may only have a few minutes to **escape** (es-KAYP). If you live near a coast, move to higher ground right away if you feel an earthquake. Don't return home until the experts say it's safe.

Tsunamis don't happen very often. Hawaii and Japan have one or two per year. The states on the west coast of the United States, such as Alaska and Washington, have one every fifteen to eighteen years.

The Future

As many as two local tsunamis happen every year. And if experts can **predict** (pre-DIKT) tsunamis, they can warn people early, and give them more time to escape.

Once a tsunami has formed, experts know when it will reach land. But scientists are trying to find a way to tell if a tsunami is going to form—even before it has. They use instruments to record movements on the ocean floor. By recording movements they hope to learn more about these giant waves. The more scientists know about tsunamis, the safer people around the world will be.

Web Sites:

You can learn more about tsunamis at this Web site:
http://www.fema.gov/kids/

Glossary

crust (KRUST) The outer layer of Earth; it is broken up into many pieces.

damage (DAM-ij) Harm or injury.

debris (duh-BREE) Pieces of wood, glass, stone, and other materials left after a disaster.

destroy (dih-STROY) To tear apart or ruin.

disturbance (dih-STER-bints) A sudden movement that upsets something that is usually still.

earthquake (ERTH-kwayk) When the crust of Earth shakes as a result of two plates running into one another.

epicenter (EH-pih-sen-ter) The place on the ground right above where an earthquake happens.

eruption (ee-RUP-shun) The spewing of gases, ash, and molten rock from a crack in Earth's crust.

escape (es-KAYP) To get away and avoid harm or injury.

evacuate (ee-VA-kyoo-ayt) To leave an area immediately because it is not safe.

expert (EK-spert) A person who has a lot of knowledge about one subject.

force (FORS) The power of something.

maximum (MAK-sih-mum) The most.

meteorite (MEE-tee-or-yt) A rock from space that has survived the trip through Earth's atmosphere and lands on Earth.

plate (PLAYT) A moving piece that makes up Earth's crust.

predict (pre-DIKT) To say something will happen before it does.

seismograph (SYZ-muh-graf) An instrument that measures movement in Earth's crust.

shallow (SHA-loh) Not deep.

tide (TYD) The rise and fall of the surface of the ocean.

tsunami (soo-NAH-mee) A series of waves caused by a disturbance in Earth's crust in or near the ocean.

wavelength (WAYV-lenkth) The distance between two waves in a series of waves.

Index